P9-DEQ-985

Ms. Tosh holds up her field guide for the class to see. "This field guide is a book about insects that live near the pond," she says. "It has pictures and information about the insects and their habitats. The field guide includes measurements. It shows the size of many insects. It has information about the plants in the area, too."

Ms. Tosh says students will work in pairs to find and identify the insects. Each pair will use a field guide.

Students will use field guides to learn about the insects they see on their field trip.

FIELD GUIDE

Hickory Horned Devil
About 10 cm long

Chapter 2

At the Pond

Today is the field trip. The students brought their science journals to record what they see.

Ms. Tosh has field guides. She packed colored pencils so the students can draw pictures of insects they see. Ms. Tosh says the students will look for colors and markings to help them identify the insects. They will estimate the sizes of the insects. They will compare the estimates with the sizes in the field guide.

The bus arrives at the pond. Students put on gloves before starting their investigations.

A Trip to the Pond

by Linda Bussell

Copyright © Gareth Stevens, Inc. All rights reserved.

Developed for Harcourt, Inc., by Gareth Stevens, Inc. This edition published by Harcourt, Inc., by agreement with Gareth Stevens, Inc. No part of this publication may be reproduced or transmitted in any form or by any means, electronic or mechanical, including photocopy, recording, or any information storage and retrieval system, without permission in writing from the copyright holder.

Requests for permission to make copies of any part of the work should be addressed to Permissions Department, Gareth Stevens, Inc., 330 West Olive Street, Suite 100, Milwaukee, Wisconsin 53212. Fax: 414-332-3567.

HARCOURT and the Harcourt Logo are trademarks of Harcourt, Inc., registered in the United States of America and/or other jurisdictions.

Printed in China

ISBN 13: 978-0-15-360242-9
ISBN 10: 0-15-360242-2

8 9 10 0940 16 15 14 13
4500409952

Harcourt
SCHOOL PUBLISHERS

Chapter 1

A Field Trip

The classroom buzzes with news. The students in Ms. Tosh's class are going on a field trip!

Ms. Tosh says they will visit the local pond and the area around it. The students will look for different plants and insects. They will observe the insects in their habitats. The habitats include the pond, the soil, milkweed plants, other wildflowers, and nearby trees.

They will use their science journals to record what they see. Later, students will write reports about their findings to share with the class.

Milkweed is a source of food for many different insects.

Milkweed and other wildflowers grow near the pond. They also grow in the field around the pond. These plants are an important habitat for some insects. This milkweed is about 90 cm tall. That is more than the length of an adult baseball bat!

The students notice many orange and black butterflies flying around the milkweed. At first all the butterflies look the same. A few are different though. Adam and Rachel use their field guide to discover that there are two kinds of orange and black butterflies.

One of the butterflies is called the Monarch. The other is called the Viceroy. Adam notices that the black lines on the hind wings of the butterflies are different.

They check the field guide. It says that wingspan is the distance across the widest part of the wings when they are fully open.

The Monarch's wingspan is about 10 cm. The Viceroy's wingspan is about 8 cm. Rachel records the two butterflies and their wingspans in the journal. Then Rachel sees something green hanging from a milkweed stem.

Monarch butterfly

Viceroy butterfly

A Monarch butterfly chrysalis, or pupa, is shown in different stages of development.

Adam finds a picture of the green object in the field guide. It is a Monarch chrysalis, or pupa. The Monarch caterpillar turns into a chrysalis before it becomes an adult butterfly. The process of changing from a caterpillar to a pupa to a butterfly is called *metamorphosis*.

The chrysalis in the field guide measures more than 2 cm in length. Rachel estimates this chrysalis is almost the same size as the one in the field guide. She notes this in their journal. She draws a picture of the chrysalis.

Daisy and Ruben are also exploring the milkweed. They see yellow, black, and white caterpillars. The caterpillars are different sizes. Adam reads in the field guide that they are all Monarch caterpillars.

Caterpillars grow in stages called *instars*. Between instars, the caterpillars shed their skin to keep growing. Daisy and Ruben find a table in their field guide. It compares the sizes of the Monarch instars.

Monarch butterfly instars

Stage	Approximate Length	
First Instar	About $\frac{1}{2}$ cm	
Second Instar	Almost 1 cm	
Third Instar	About 1 cm to $1\frac{1}{2}$ cm	
Fourth Instar	About $1\frac{1}{2}$ cm to $2\frac{1}{2}$ cm	
Fifth Instar	About $2\frac{1}{2}$ cm to $4\frac{1}{2}$ cm	

Daisy and Ruben compare the caterpillars they see with the information in the table. They record their observations in their journal.

A Whirligig beetle resting on the surface of a pond.

Ruben and Daisy notice movement near the edge of the pond. They see small insects whirl on the surface of the water.

Ruben finds the insect in the field guide. It is called a Whirligig beetle. The beetle is named for its whirling motion. The Whirligig beetle has unusual eyes. The eyes are divided into two parts. One part allows the beetle to see above water. The other part lets it see beneath the water.

Daisy records the Whirligig beetle in their journal. She estimates its length is about 1 cm.

Some Antlions larvae are smaller than 1 cm. Adult Antlions can have wingspans up to 1 dm.

Kami sees several small, circular pits in the sandy soil around the pond. She wonders what they could be.

Sydney finds a picture of the pits in the field guide. Antlion larvae build these pits to catch prey.

The Antlion larva in the field guide measures less than 1 cm. The adult Antlion is much larger than the larva. The adult in the field guide has a wingspan of almost 1 dm. One decimeter equals ten centimeters.

Kami records information about the Antlion larva in their journal.

Adult Luna moths can have a wingspan of more than 11 cm.

Hickory trees near the pond are another insect habitat. This type of tree can grow to be 40 m tall. One meter equals 100 centimeters. Forty meters are longer than three school buses parked end to end.

Sydney spots a large, green moth on a tree trunk. It is an adult Luna moth. This Luna moth is sitting high in the tree. It is more than a meter above their heads.

The field guide says that some adult Luna moths have a wingspan of more than 11 cm.

The Hickory Horned Devil caterpillar looks fierce, but it is harmless to people.

Benjamin and Carl also are looking among the hickory trees for insects. They are trying to find a caterpillar called the Hickory Horned Devil. It is fierce-looking, but harmless to people.

They find several Hickory Horned Devil caterpillars in the twigs of a hickory tree. The caterpillars are different sizes. They are eating hickory leaves.

Carl reads the field guide. It shows a Hickory Horned Devil that is about 10 cm long. Benjamin records the information in their science journal. Then he draws a picture of this insect.

Regal moth

Carl reads in the field guide that the Hickory Horned Devil is the caterpillar stage of the adult Regal moth. Like the Hickory Horned Devil, the Regal moth can grow very large. The distance between the two wing tips is about 10 cm.

Carl and Benjamin look around, but they do not find any Regal moths. Then they hear Ms. Tosh call the class together. She collects the pencils, markers, and field guides. Ms. Tosh collects their gloves. The students climb on board the bus.

Chapter 3

Going Home

The students settle on the bus. They talk all at once about the different insects they saw at the pond. Ms. Tosh asks them to name some of the insects they saw. The students talk about some of the surprising things they learned.

"We saw lots of Monarch butterflies," says Rachel. "We read that some Monarchs fly more than 4,000 km!"

"Correct," says Ms. Tosh. "Some Monarchs migrate from southern Canada, across the United States, to central Mexico. That is a very long trip for such a small insect!"

The students saw many insects at the pond.

Ms. Tosh asks her students to share some of their drawings. The students show the pictures they made. They show drawings of butterflies, dragonflies, mantids, caterpillars, and even a giant water bug. The students also discuss the journal entries they made at the pond.

"You have learned a lot today," says Ms. Tosh. "I am proud of you. I cannot wait to read your reports."

It has been a busy day. This has been a fun field trip. The bus driver starts the bus, and they head back to school.

Glossary

centimeter (cm) a metric unit that is used to measure length or distance. 100 centimeters = 1 meter

chrysalis the pupa of a butterfly

decimeter (dm) a metric unit that is used to measure length or distance. 1 decimeter = 10 centimeters

habitat the place or environment where a plant or animal naturally lives and grows

insect class of small animals. Insects have 6 legs and 3 main body parts called the head, thorax, and abdomen. Many insects have one or two pairs of wings.

investigation a close study of something to find out information

kilometer (km) a metric unit that is used to measure length or distance. 1 kilometer = 1,000 meters

larva an insect in its early life stage, between hatching from an egg and becoming a pupa. *Larvae* is plural for *larva*.

meter (m) a metric unit that is used to measure length or distance. 1 meter = 100 centimeters.

wingspan the distance across the widest part of the wings when they are fully open

Photo credits: cover, title page, p. 15 (left)U.S. Fish and Wildlife Service; pp. 3, 7 (main image): © S & D & K Maslowsk/FLPA; pp. 5, 6 (both), 7 (inset), 8, 9 (inset), 10 (inset), 11, 12, 13: Totallybuggin.com; pp. 9 (main image), 10 (top): © Foto Natura Stock/ FLPA; pp. 10 (bottom), 15 (right) © Corbis; p. 15 (center) U.S. Department of Agriculture